SATELLITES

SATELLITES

BY ALICE FIELDS
ILLUSTRATED BY MIKE TREGENZA

A GROLIER COMPANY

FRANKLIN WATTS
NEW YORK/LONDON/TORONTO/SYDNEY/1981
AN EASY-READ FACT BOOK

Thanks are due to the following for
kind permission to reproduce photographs:
British Aerospace; NASA; Novosti; Post Office
of Great Britain; Woodmansterne/NASA

Library of Congress Cataloging in Publication Data
Fields, Alice.
    Satellites.
    (An Easy-read fact book)
    Includes index.
    SUMMARY: Describes artificial satellites—what they
are, how they work, and what they can do.
    1. Artificial satellites—Juvenile literature.
[1. Artificial satellites]   I. Tregenza, Michael.
II. Title.
TL796.F54   629.46   80-13700
ISBN 0-531-03246-9

R.L.3.1 Spache Revised Formula

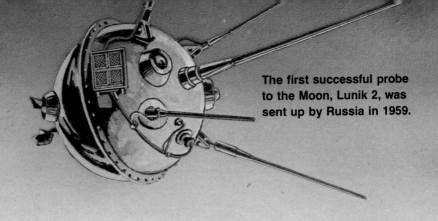

The first successful probe to the Moon, Lunik 2, was sent up by Russia in 1959.

In October 1957 a second "moon" appeared in the sky.

It was a **satellite** (SAT-el-light) that had been made and put into space by the Russians. They gave this moon, or satellite, the name **Sputnik.**

Sputnik marked the start of our Space Age. Since 1957, many satellites have been launched around Earth.

Satellites are very useful. Some help scientists to learn more about Earth and space. Some help us to forecast the weather.

Other satellites carry radio and television signals around the world.

Satellites that we make are called **artificial** (man-made) **satellites**. They are different from **natural satellites**.

The Moon is a natural satellite of Earth.

In the Sun's family, or **solar system**, six other planets have natural satellites:

Mars has two natural satellites.

Jupiter has fourteen.

Saturn has ten, possibly eleven.

Uranus has five.

Neptune has two.

Pluto has one, recently discovered.

Most natural satellites are probably great balls of rock, like the Moon.

Artificial satellites are much smaller than natural satellites. They are made mainly of metal.

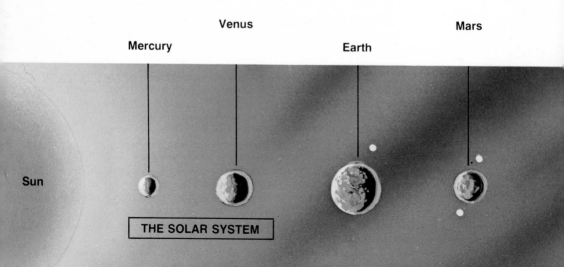

Venus

Mercury

Mars

Earth

Sun

THE SOLAR SYSTEM

A spacecraft flying over the Moon.
Earth can be seen in the background.

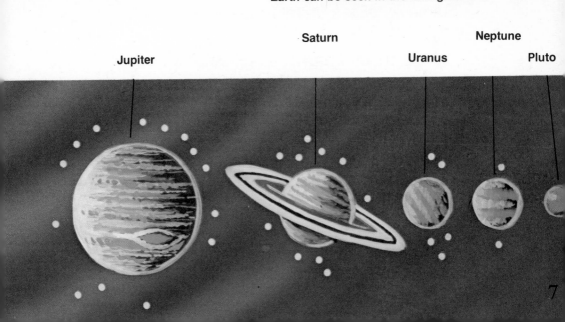

Jupiter   Saturn   Uranus   Neptune   Pluto

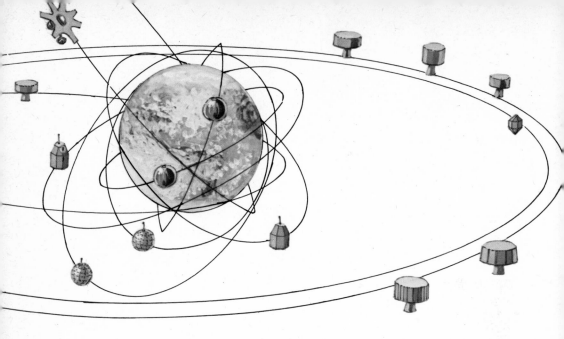

The orbits of some of the major satellites

Like the Moon, artificial satellites travel around Earth. Each satellite has a set path in space. We call this path the **orbit** (OR-bit).

Some satellites travel in an orbit that is almost a circle.

Others travel in an oval, or **elliptical** (e-LIP-t-kal) orbit. Their distance above Earth changes as they travel through space.

One Russian satellite has an orbit that takes it very close to Earth and then very far away. It comes as close as 310 miles (500 km). At its farthest, it is 24,000 miles (39,000 km) away!

Satellites travel at very high speeds. They go faster than the fastest jet airplane. Therefore, they must travel above Earth's **atmosphere** (AT-mo-sfear).

If a satellite traveled through the atmosphere, air would push against it and slow it down. And air rubbing against it would produce tremendous heat. The satellite might burn up.

To be above the atmosphere, satellites must travel at least 180 miles (300 km) above Earth.

Stages of a satellite burning up in the atmosphere

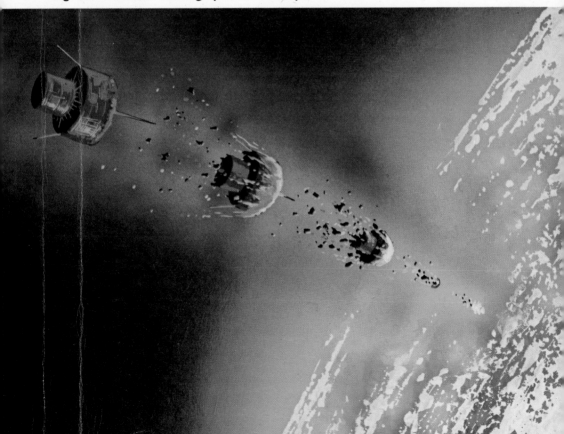

It may be surprising that a satellite stays out in space and does not fall back to Earth.

Usually Earth's pull, or **gravity** (GRAV-it-e), makes everything fall to the ground. But a satellite overcomes gravity by speed.

To understand this, imagine a gun firing shells from a mountain. The gun can fire shells at different speeds.

As soon as a shell is fired, gravity starts to pull it to the ground. The faster the shell travels, the father it will go before gravity pulls it down.

Overcoming the force of gravity (represented by red arrows)

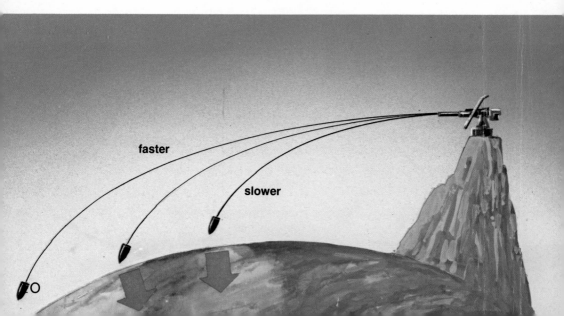

The shell at first overcomes gravity by speed. Now suppose the shell travels at a speed of 17,500 miles (28,000 km) per hour. And suppose the mountain is 1,980 miles (3,186 km) high.

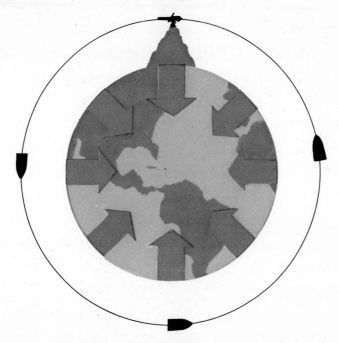

A shell overcoming the force of gravity (represented by red arrows)

At this height and speed, the shell would never hit the ground!

As the shell falls, Earth curves away from under it at exactly the same speed. So the shell always stays the same height above the ground.

The same thing happens with a satellite.

To get a satellite into orbit, then, two things must be done.

The satellite must be pushed beyond the atmosphere. And it must have a speed of at least 17,500 miles (28,000 km) per hour.

To do this, a very powerful engine is needed. The engine must be able to work in space, without air.

The only engine that can do this is a **rocket**.

A rocket burns fuel inside it. This produces a stream of hot gasses. The gasses shoot out the back, pushing the rocket forward.

But one rocket is not yet strong enough to put a satellite into orbit. Two or more rockets must be joined together.

The rockets are placed on top of one another. Together, they form a **multi-stage rocket**.

Each rocket engine gives the ones above it a "piggyback" ride into space. Each part, or **stage**, fires in turn. Then it falls away when its fuel has been used up.

Finally, just the satellite is left in space.

THIRD STAGE

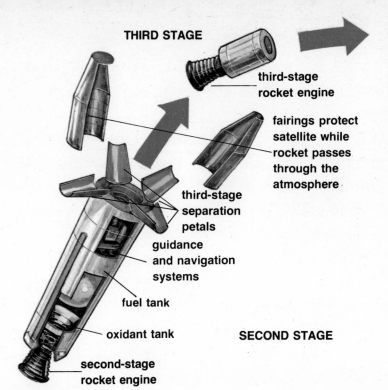

third-stage
rocket engine

inflatable satellite
("Echo") folded
inside container

fairings protect
satellite while
rocket passes
through the
atmosphere

third-stage
separation
petals

guidance
and navigation
systems

fuel tank

oxidant tank

SECOND STAGE

second-stage
rocket engine

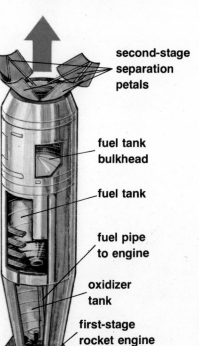

second-stage
separation
petals

**LAUNCHING A SATELLITE
USING A MULTI-STAGE ROCKET**

fuel tank
bulkhead

fuel tank

FIRST STAGE

fuel pipe
to engine

oxidizer
tank

first-stage
rocket engine

stabilizer
fin

13

Now let's take a look at some different kinds of satellites.

OAO is an Orbiting Astronomical Observatory. It collects information about the stars. Then it radios that information back to Earth.

OAO contains different measuring instruments and several telescopes.

The instruments and radio work by electricity. The electricity is produced in the "wings" of the satellite.

The wings are called **solar panels.** Solar panels are made up of hundreds of tiny **solar cells**. Solar cells produce electricity from sunlight.

Most satellites have solar panels.

Orbiting Astronomical Observatory (OAO)

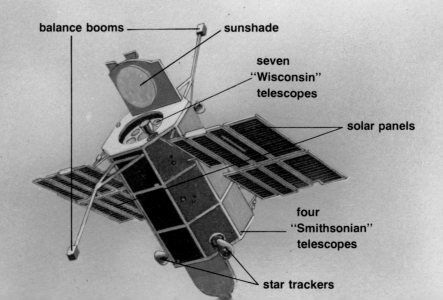

balance booms — sunshade

seven "Wisconsin" telescopes

solar panels

four "Smithsonian" telescopes

star trackers

Satellite photograph of San Francisco Bay

Landsat is another kind of satellite. It takes
photographs of Earth from space. Its purpose is to
study forests, oceans, and the health of crops.

Landsat can even spot diseased crops or forest
fires.

Photographs taken of Earth from space are very important to **cartographers** (car-TOG-ra-fers). Cartographers are people who make maps.

The Grand Canyon area photographed by satellite

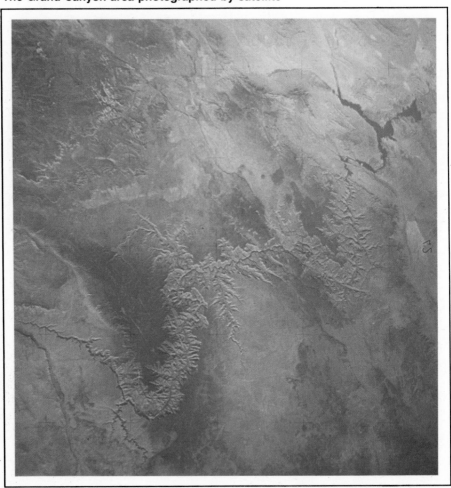

It is interesting to study photographs of Earth taken from space. Compare them with maps in an atlas. You can see if the maps are correct.

Satellite photograph of a section of the Gulf of California, showing the mouth of the Colorado River

Spiral storm clouds

Weather satellites are very useful. They constantly take pictures of cloud formations over Earth.

These pictures can tell **weather forecasters** when a storm is developing. Forecasters can see where the storm is and where it is heading.

If a storm looks as though it will become violent and dangerous, people can be warned. Advance warning may save hundreds of lives.

Tiros, the world's first weather satellite, was launched in 1960. Solar cells are around the body, rather than on wings.

Nimbus "E", launched in 1972, has solar cells on wings. The advantage is that wings can be turned to face the sun as long as possible.

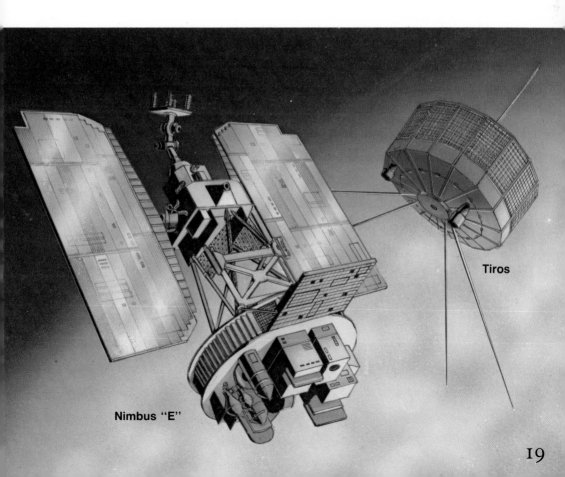

Tiros

Nimbus "E"

A weather satellite usually circles at about 560 miles (900 km) above Earth. Its orbit takes it over the North and South Poles.

A weather satellite orbits Earth about every 100 minutes.

During this time, Earth itself is turning. Therefore, on the next orbit, the satellite goes over a slightly different part of Earth.

As Earth turns below it, the satellite can finally scan the whole globe. Then all the pictures are put together.

In this way, a whole picture of the cloud cover over Earth can be formed.

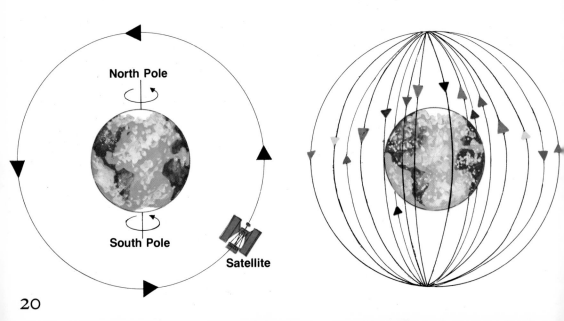

Photograph of Earth taken by the Apollo astronauts when they were 106,000 miles (170,600 km) out in space. The white areas are clouds.

Space scientists can send signals to a satellite at any time. They also keep receiving information back from it.

This way, they can keep careful track of an orbiting satellite and know where to send the signals.

They usually track (trace) satellites by using **radio waves**. The radio waves, or **signals**, are picked up by large **antennae** (an-TEN-i).

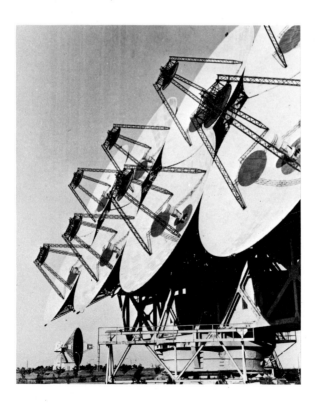

Russia's tracking station in the Crimea

A tracking ship

Both the United States and Russia have wide networks to track satellites. They have **tracking stations** and **tracking ships** around the world.

The stations have large dish-shaped antennae. Some are 200 feet (about 60 m) across.

These huge antennae can pick up weak radio signals from a distant satellite. They also send out strong beams of signals themselves.

Signals from the ground can be aimed exactly at a satellite far out in space.

A satellite collects information as it travels through space. But it does not send the information back to Earth right away.

There may not be a tracking station within range. So the satellite first records the information on a tape recorder.

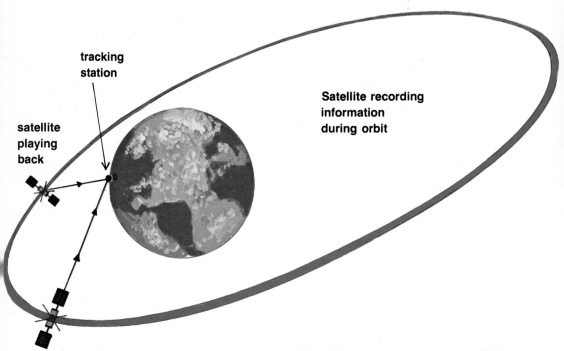

tracking
station

satellite
playing
back

Satellite recording
information
during orbit

Sooner or later a ground (Earth) station comes within range.

The ground station sends out a signal telling the satellite to play back the tape. The information is then played back over the satellite's radio.

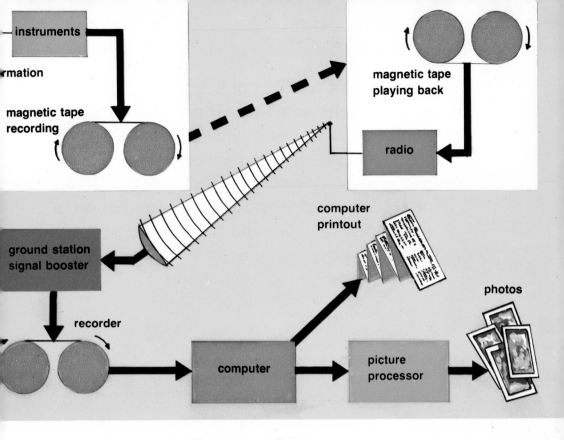

instruments

rmation

magnetic tape
recording

magnetic tape
playing back

radio

ground station
signal booster

computer
printout

photos

recorder

computer

picture
processor

Just before the satellite goes out of range, the ground
station orders it to stop transmitting.

At the ground station, the signals from the satellite
are **processed**. (A computer prints out the
information.)

Certain signals are translated into written facts and
figures. Light signals are processed into
photographs.

The most useful satellites of all are the **communications satellites.** They are used to **relay** (pass) signals from one ground station to another.

They relay radio signals, television programs, and telephone messages over great distances.

The first communications satellite was **Echo I**, a silvery balloon 100 feet (about 30 m) across.

It had no special sending or receiving instruments. Echo I was a **passive satellite**.

Echo I

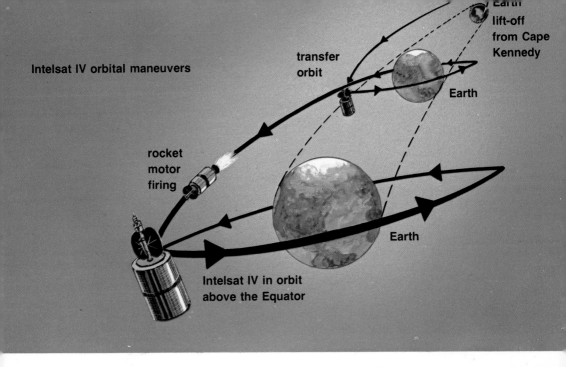

Intelsat IV orbital maneuvers

transfer orbit

Earth lift-off from Cape Kennedy

Earth

rocket motor firing

Earth

Intelsat IV in orbit above the Equator

Signals from a ground station were reflected off Echo's shining surface. They simply bounced back to another ground station.

But these signals were very weak by the time they reached the receiving station. A satellite that could **amplify** (AM-pla-fi), or strengthen, the signals was needed.

Today, most satellites are **active satellites**. They can strengthen the signals they receive. Then they beam the strong signals back to Earth.

Several communications satellites are now orbiting Earth. The pictures on the next two pages show a **communications satellite system**.

UNITED STATES

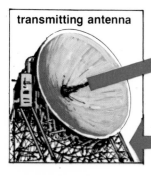

transmitting antenna

communication
satellite over the
Atlantic Ocean

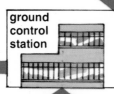

ground
control
station

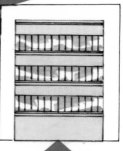

telephone
exchange

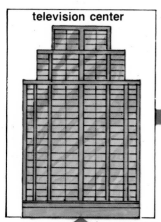

television center

television camera
filming motor race

outgoing
telephone call

28

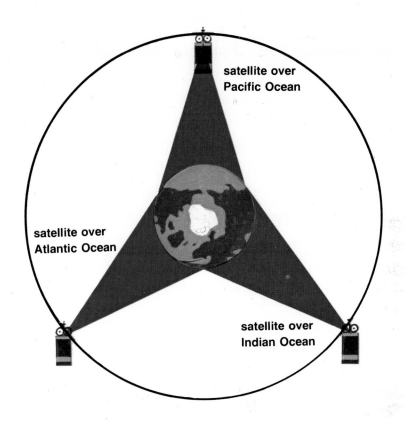

satellite over
Pacific Ocean

satellite over
Atlantic Ocean

satellite over
Indian Ocean

But Earth itself turns completely around in 24 hours.

Thus, the satellite keeps exactly in step with Earth. It appears from Earth to be standing still.

The satellite seems stationary in the sky, but it is really moving.

Three satellites in stationary orbit can be spaced to serve nearly the whole world.

In the United States, a private company called COMSAT owns and operates commercial communications satellites. The ground station is in Andover, Maine.

In Great Britain, the Government Post Office (GPO) controls communication by satellite. Their ground station is at Goonhilly Downs, in Cornwall.

INTELSAT is a world-wide satellite system. The member nations jointly own the satellites. The ground stations are owned by the government in some countries, by private companies in others.

One of the antennae at Goonhilly Downs

Intelsat IV

There are other communications satellite systems, too. For example, there are systems for defense and military communications.

A special kind of temporary satellite is the manned spacecraft. It carries **astronauts** (ASS-tro-nawts) into space.

**An astronaut from Apollo 11
traveling on the Moon in a lunar rover**

Manned spacecraft are much bigger and much more complicated than instrument satellites.

Usually, a manned spacecraft stays in orbit for a few days. But it is equipped to stay up for several weeks. Then it returns to Earth.

Manned spacecraft must carry large supplies of water and food. They must even carry the air the astronauts need to breathe in airless space.

These spacecraft are equipped with computers, rocket motors, and fuel. They must carry instruments for the astronauts to work with in space.

The astronauts also need special equipment to bring them safely back to Earth.

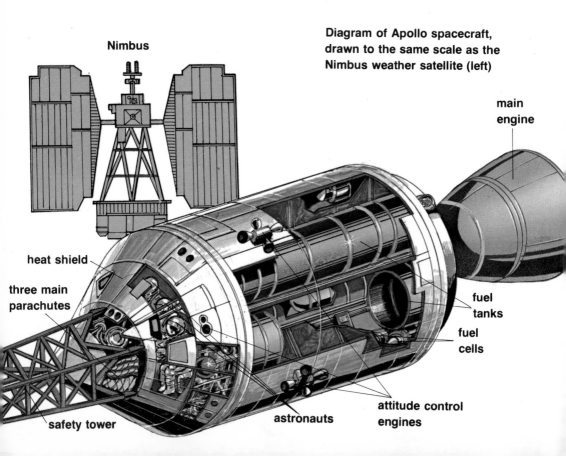

Nimbus

Diagram of Apollo spacecraft, drawn to the same scale as the Nimbus weather satellite (left)

main engine

heat shield

three main parachutes

fuel tanks

fuel cells

attitude control engines

astronauts

safety tower

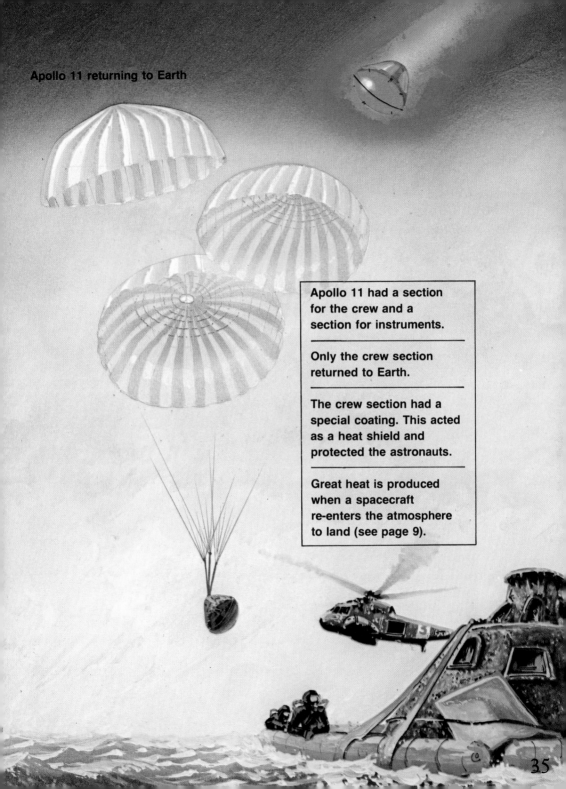

Apollo 11 returning to Earth

Apollo 11 had a section
for the crew and a
section for instruments.

Only the crew section
returned to Earth.

The crew section had a
special coating. This acted
as a heat shield and
protected the astronauts.

Great heat is produced
when a spacecraft
re-enters the atmosphere
to land (see page 9).

35

In the future, the United States will launch its astronauts in a **space shuttle**. It will be controlled by a pilot inside the spacecraft.

The shuttle can be launched into orbit over and over again. Former spacecraft could be used only once.

The space shuttle will cost much less to operate over a period of time.

Launch of space shuttle

Orbiter in orbit

The main part of the space shuttle is called an **orbiter** (OR-bit-er). It has wings like a jet airliner and is about the same size.

Most of the engine fuel is carried in a huge tank outside the orbiter. Two **booster rockets** are attached to the tank.

The orbiter's engines and the booster rockets fire together. This provides enough power to send the orbiter into space.

When the boosters run out of fuel, they drop off. Parachutes (PAHR-a-shoots) float them back to Earth.

The engines of the orbiter keep firing until it is in orbit. Then the empty outside fuel tank is dumped.

The shuttle orbiter has a huge space for cargo. It is called the **cargo bay**.

The orbiter's cargo bay can hold 30 tons of equipment! It can carry several satellites into orbit at once.

Astronauts from the shuttle can check over satellites already in orbit. They can make repairs out in space. Or they can carry satellites back to Earth, if necessary.

Another important cargo is **spacelab**. This is an entire laboratory unit. It is made to fit into the shuttle's cargo bay.

Launching satellites from a space shuttle

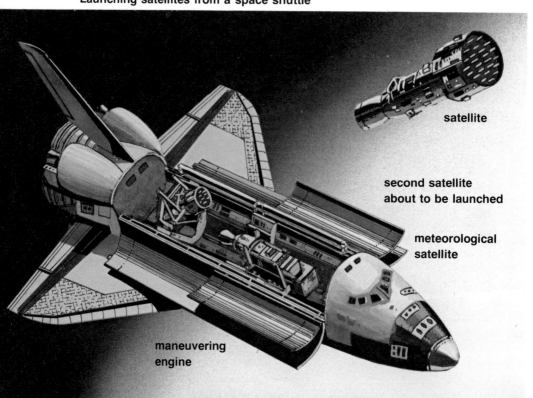

satellite

second satellite
about to be launched

meteorological
satellite

maneuvering
engine

When the orbiter is ready, it can return to Earth.

The orbiter crew fires **retro rockets**. These slow the shuttle down so that it falls toward Earth. It slows down even more as it re-enters the atmosphere.

Once in the atmosphere, the orbiter can be flown like a glider. The pilot lands it on a specially-built runway.

In a few weeks, the orbiter can be made ready for another flight.

Orbiter landing

Space scientists are very interested in building **space stations**. These would be permanent, manned satellites.

In 1971, the Russians launched a space station called **Salyut** (SOL-yet). It carried no people—only instruments.

Then a team of three **cosmonauts** (Russian astronauts) went up in a **Soyuz** (SOY-ez) spacecraft. They were able to link up with Salyut.

The three cosmonauts stayed on Salyut for over three weeks. They carried out many experiments there.

But during their return trip to Earth, something went wrong and the three cosmonauts were tragically killed. They have become Russian heroes. They showed the world that an orbiting space station was possible and useful.

Since then, Russia has launched a number of Salyut space stations. In 1978, two cosmonauts spent a record 120 days in orbit in Salyut 6.

While in orbit, their Salyut station received supplies from automatic cargo ships. And they were even visited by other cosmonauts!

**Soyuz (bottom) about to join up
with Salyut space station (top)**

Skylab was America's first experimental space station. It was launched in 1973.

Three crews of astronauts visited it at different times. The astronauts used simplified Apollo spacecraft to travel to and from Skylab. The most recent crew spent 84 days there in 1973.

Interior diagram of Skylab

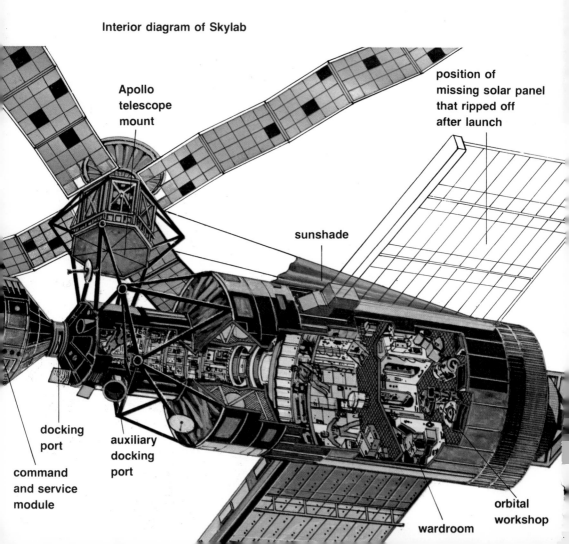

Apollo telescope mount

position of missing solar panel that ripped off after launch

sunshade

docking port

auxiliary docking port

command and service module

orbital workshop

wardroom

Possible space station of the future

In Skylab, the astronauts carried out hundreds of scientific experiments. They were studying the weightless and airless conditions of space.

They also used special telescopes to view the stars. Being above Earth's atmosphere, their view was perfectly clear.

The United States hopes to build much bigger space stations. They could remain in orbit for many years.

The parts needed to build bigger space stations would be carried by space shuttle. Then the stations would be put together out in space.

As we have seen, some spacecraft are artificial satellites that orbit Earth.

Scientists also send out spacecraft that are not meant to orbit Earth. They are called **probes**.

Probes go millions of miles into space. They are sent to investigate other planets.

Many probes were sent to the Moon before the Apollo astronauts landed there.

The **Lunar Orbiter Probes** became artificial satellites of the Moon itself. They photographed the surface of the Moon very clearly.

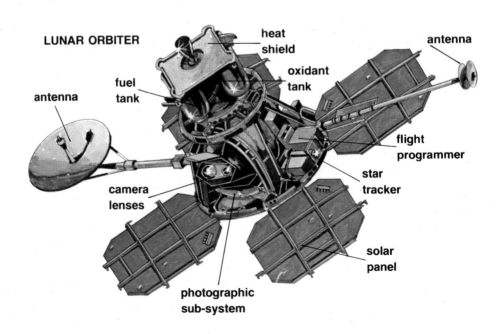

LUNAR ORBITER

heat shield

antenna

oxidant tank

fuel tank

antenna

flight programmer

star tracker

camera lenses

solar panel

photographic sub-system

The photograph above shows the Moon crater
Tycho. It is about 50 miles (80 km) across.

Several probes have been sent to the planet Mars. Some have become satellites of Mars.

The American probe, **Mariner 9**, became the first artificial satellite of Mars in 1971.

In eleven months, Mariner 9 sent back more than 7,000 photographs. They showed the whole surface of the planet Mars. There were pictures of deep craters. There were also pictures of some volcanoes that might still be active.

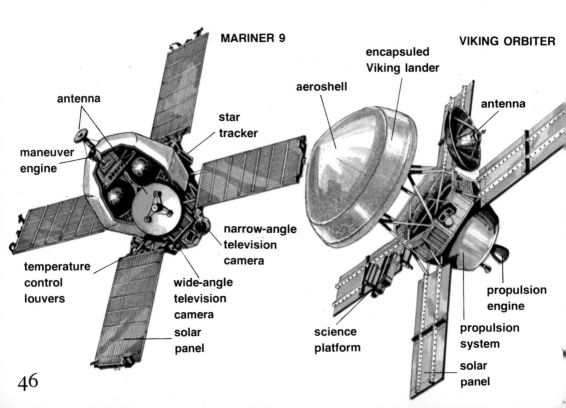

MARINER 9

antenna

maneuver engine

star tracker

temperature control louvers

narrow-angle television camera

wide-angle television camera

solar panel

aeroshell

encapsuled Viking lander

VIKING ORBITER

antenna

science platform

propulsion engine

propulsion system

solar panel

In 1975, Mars got two new satellites, **Viking 1** and **2**.

The Viking probes sent out landing craft. These took close-up pictures of the surface of Mars. They showed the orange-red soil that gives Mars its fiery color.

Picture of the surface of Mars taken by Viking. An arm of the Viking probe can be seen on the right of the picture.

Scientists have only just started to discover the mysteries in space. We can only guess what the future may reveal!

INDEX